se
pup

penguin
chick

Look and See

Can you

see my

black stripes?

2 COGNITIVE DEVELOPMENT

Can you see my white spots?

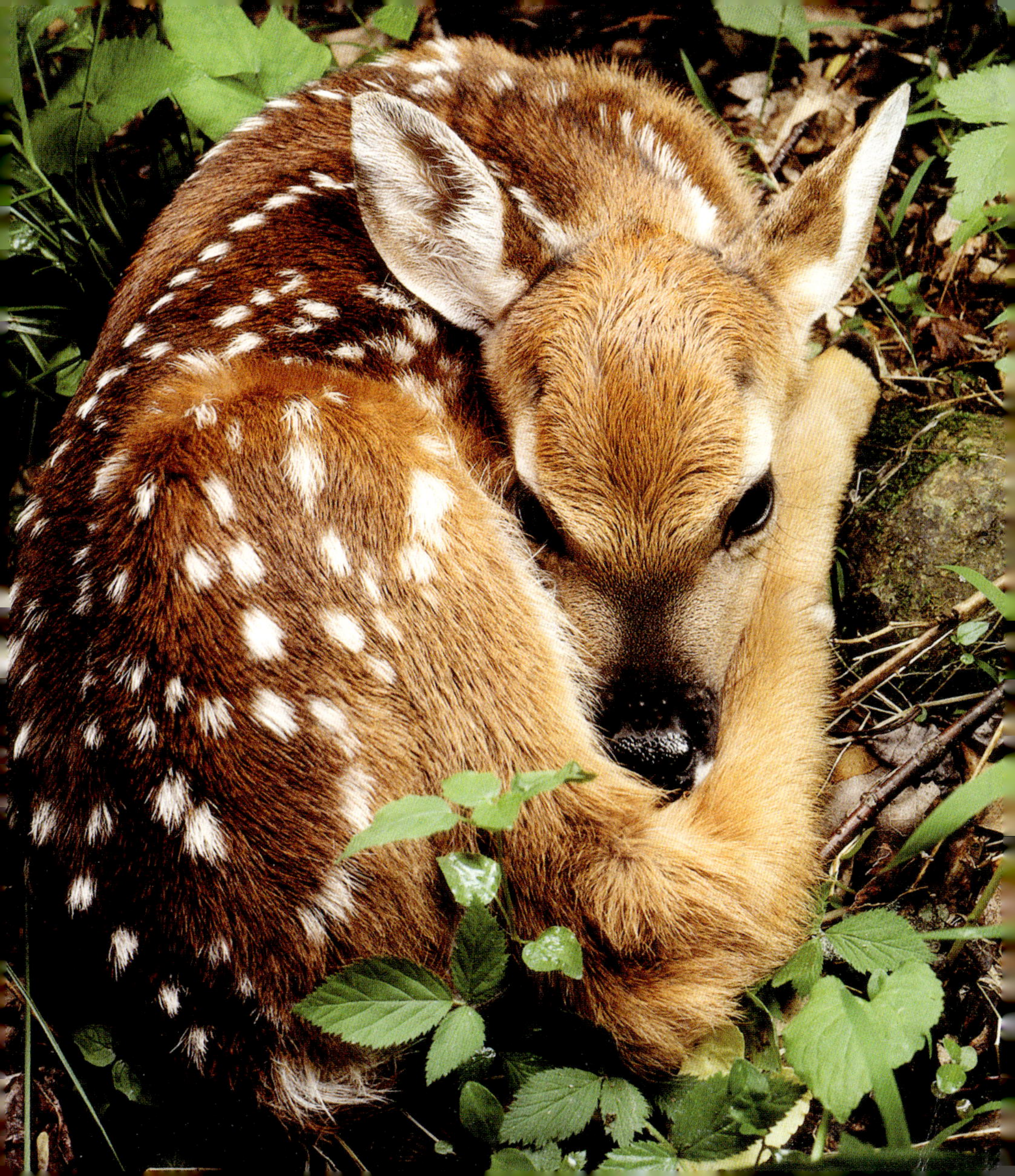

Peek-A-Boo

Take a peek. What do you see?

LANGUAGE DEVELOPMENT

Head-to-Toe

Silly, hap

open here

8 SOCIAL/EMOTIONAL DEVELOPMENT

LEMUR

A baby on a back looking at me.

A baby in a pouch looking at me.

WALLABY

py   faces.

Busy, **lit**

tle feet.

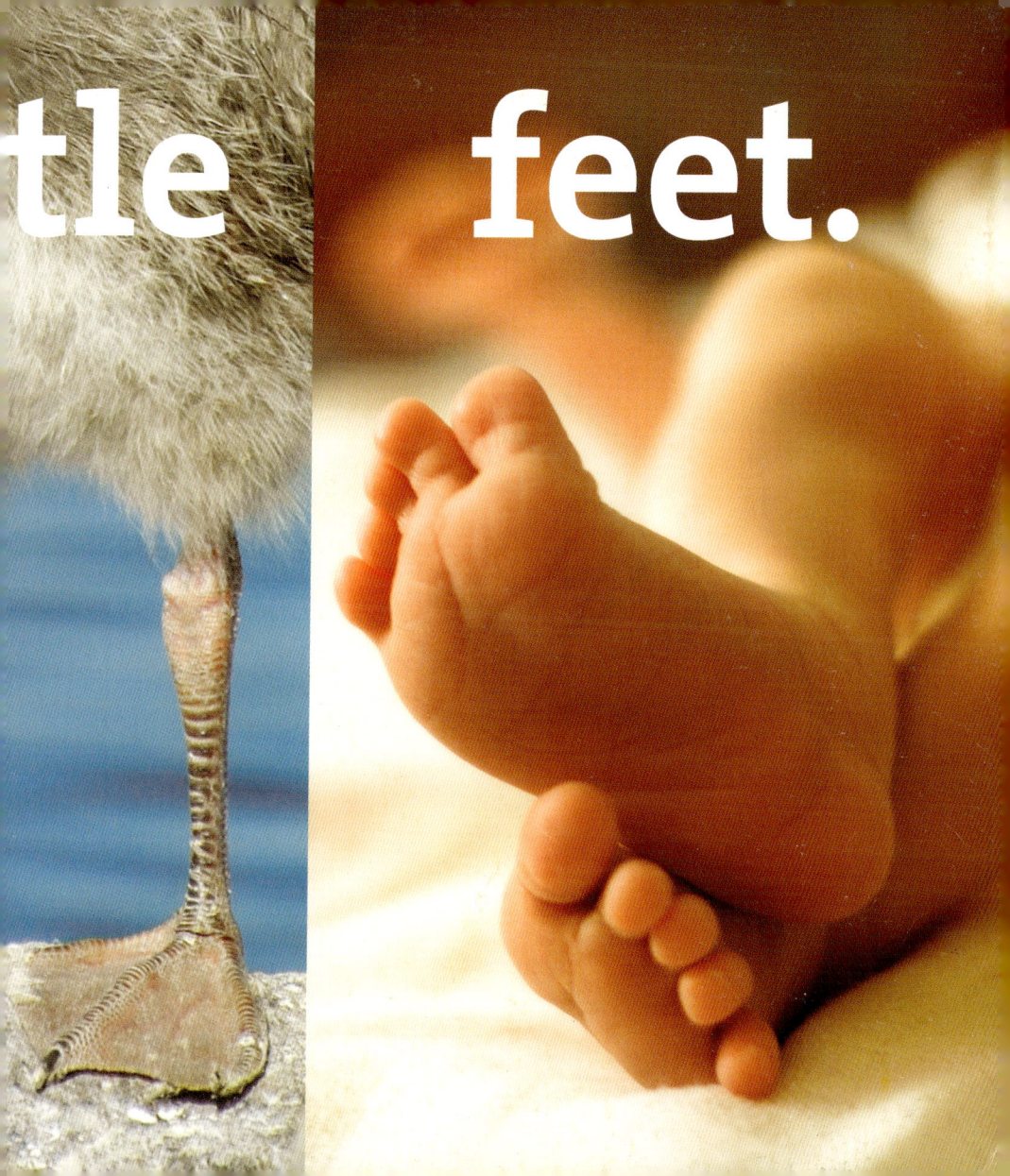

Just Like Me!

Jump, jump, jumping.

SOCIAL/EMOTIONAL DEVELOPMENT

It's time to **P L A Y**.

Munch, munch, munching.

It's time to EAT.

Nap, nap, napping.

It's time to SLEEP.

Rhyme Time

Three baby rabbits
    Happy as can be,
Hopping in the green grass
    One, two, three.

Zoombies + You

PLAYTIME!

Spontaneous play filled with laughter is an important way for us to connect with our children. Find a small handheld mirror. Hold the mirror up so both you and your child can easily see your faces. It's easy to make all sorts of silly, happy faces. Raise your eyebrows. Purse your lips. Stick out your tongue. Let the giggling begin!

GREEN TIP

Making our own toddler food can be a healthy alternative. If possible, buy organic fruit to make simple purees. Consider combining all sorts of flavors and textures, such as mango, banana, and strawberries. Carefully wash fruits before making the puree. Peel and cut fruits into bite-size pieces. If necessary, steam or bake hard fruits. Place the fruit in a blender or food processor. If there's any extra puree, place it in an ice cube tray to freeze and eat later as an ice pop.